Collins

T0318024

CSEC® MATHS
WORKBOOK

Contents List

Terry David

Collins

HarperCollins Publishers Ltd
The News Building
1 London Bridge Street
London SE1 9GF

HarperCollins Publishers Ltd
Macken House,
39/40 Mayor Street Upper,
Dublin 1,
D01 C9W8,
Ireland

First edition 2015

10 9 8 7 6

ISBN 978-0-00-814739-6

Collins® is a registered trademark of HarperCollins Publishers Limited

Maths Workbook for CSEC is an independent publication and has not been authorised, sponsored or otherwise approved by **CXC**®.

CSEC® is a registered trade mark of the **Caribbean Examinations Council (CXC)**.

www.collins.co.uk/caribbeanschools

A catalogue record for this book is available from the British Library.

Typeset by QBS
Printed by Ashford Colour Press Ltd

If any copyright holders have been omitted, please contact the Publisher who will make the necessary arrangements at the first opportunity.

Author: Terry David
Illustrators: QBS
Publisher: Elaine Higgleton
Commissioning Editor: Peter Dennis
Managing Editor: Sarah Thomas
Editor: Tanya Solomons
Proofreader: Lucy Hyde
Technical Reviewer: Amanda Dickson

1 ▶ Computation

1 Convert EACH of the following to a decimal.

a) $\frac{1}{8}$ [2]

b) $\frac{2}{3}$ [2]

c) $\frac{1}{5}$ [2]

2 Convert EACH of the following to a fraction.

a) 0.375 [2]

b) 0.75 [2]

c) 0.02

[2]

3 Convert EACH of the following to a percentage.

a) $\frac{5}{8}$

[2]

b) 0.135

[2]

c) $\frac{5}{7}$

[2]

4 **a)** Express 3 768 to 3 significant figures.

[1]

b) Express 1.068 to 3 significant figures.

[1]

c) Express 1.201 to 2 significant figures.

[1]

d) Express 25 323 to 2 significant figures.

[1]

5 Write EACH of the following in standard form.

a) 2 500 000 [1]

b) 0.003 251 [1]

c) 362 000 [1]

d) 0.000 009 [1]

6 a) The angles of a triangle are A, B and C. The angles are in the ratio 2 : 3 : 1. Calculate the values of A, B and C. [3]

b) A rope has a length of 40 m. It is cut into pieces in a ratio of 2 : 3 : 5. Calculate the length of the longest piece. [2]

c) A sum of money is divided between Melissa and Kerry in a ratio of 4 : 1. Kerry receives $45. How much money was shared in total? [2]

7 A bus leaves Sangre Grande in Trinidad at 7:25 a.m. It travels a distance of 92 km and arrives in San Fernando at 9:30 a.m.

a) Calculate the duration of the journey. [2]

b) Calculate the average speed for the entire journey. [2]

8 Calculate:

a) the exact value of $\dfrac{2\frac{1}{2} \times \frac{3}{5}}{\frac{1}{2} - \frac{1}{5}}$ [4]

b) the value of $14.25 - (1.24)^2$ correct to 3 significant figures [3]

9 Evaluate the following:

a) 2.14 (3 − 1.26) [2]

b) $\dfrac{2.15}{0.8^2 - 0.22}$ [3]

10 Jesse is trying to estimate the distance between two cities on a map. The scale on the map is 1 : 10 000 000. The distance between the two cities on the map is 4.2 cm. Calculate the distance between the two cities in km. [2]

11 The scale on a map of Jamaica is 1 : 20 000. Complete EACH of the following statements.

a) 1 cm on the map represents _____ cm on the island of Jamaica. [1]

b) 1 cm on the map represents _____ km on the island of Jamaica. [1]

c) 4.5 cm on the map represents _____ km on the island of Jamaica. [1]

12 a) What is 12.5% of 80? [1]

b) 20% of a number is 25. What is the number? [1]

c) Express 12 as a percentage of 60. [1]

d) What fraction of 50 is 10? [1]

13 a) Convert 2.5 km to m. [1]

b) Convert 3000 cm to km. [1]

c) Convert 2 litres to ml. [1]

Total Marks _____ / 61

2 Number theory

1 The sets A, B and C are defined as follows:

A = {1, 2, 3, 4, …}

B = {0, 1, 2, 3, 4, …}

C = {…, −3, −2, −1, 0, 1, 2, 3, …}

Match the sets with the following:

a) Whole numbers: _____ [1]

b) Integers: _____ [1]

c) Natural numbers: _____ [1]

2 State the factors of EACH of the following:

a) 12_____ [2]

b) 10_____ [2]

c) 21_____ [2]

3 Determine the Highest Common Factor (HCF) of EACH of the following:

a) 12 and 18

factors of 12 _____ **[2]**

factors of 18_____ **[2]**

HCF_____ **[1]**

b) 30 and 15

factors of 30 _____ **[2]**

factors of 15_____ **[2]**

HCF_____ **[1]**

c) 15, 25 and 40

factors of 15 _____ **[2]**

factors of 25_____ **[2]**

factors of 40_____ **[2]**

HCF_____ **[1]**

4 State the first FOUR multiples of EACH of the following:

a) 5_____ [2]

b) 6_____ [2]

c) 12_____ [2]

5 Determine the Lowest Common Multiple (LCM) of EACH of the following:

a) 9 and 12

multiples of 9_____ [2]

multiples of 12_____ [2]

LCM _____ [1]

b) 5 and 8

multiples of 5_____ [2]

multiples of 8_____ [2]

LCM _____ [1]

c) 6 and 7

multiples of 6_____ [2]

multiples of 7_____ [2]

LCM _____ [1]

6 Determine the next THREE numbers in each of the following sequences.

a) 0, 1, 1, 2, 3, 5, [2]

b) 2, 6, 10, 14, 18, [2]

c) 1, 4, 9, 16, [2]

7 Match each of the following laws with one of the mathematical expressions below.

associative law   distributive law   commutative law

a) $a \times (a + b) = a^2 + ab$ [1]

b) $a \times (b \times c) = (a \times b) \times c$ [1]

c) $a \times b = b \times a$ [1]

8 Convert EACH of the following to base 10.

a) 11011_2 [2]

b) 232_4 [2]

c) 1242_5 [2]

9 Convert EACH of the following:

a) 15_{10} to base 2 [2]

b) 27_{10} to base 5

[2]

c) 90_{10} to base 8

[2]

10 Determine the following:

a) $110110_2 + 11001_2$

[3]

b) $1111_2 - 1001_2$

[3]

Total Marks _____ / 74

3   Consumer arithmetic

1 The cash price of a television is $3000. It is bought on a hire purchase basis.
A customer makes a deposit of $800 and then makes a monthly deposit of $300
for 10 months.

    a) Calculate the total hire purchase price.   **[2]**

    b) Calculate the amount of money that would have been saved if the television
        was bought at the cash price.   **[1]**

2 A grocer purchases 120 boxes of cereal from a wholesaler at a price of $3360.

    The grocer sells each box of cereal for $32. Calculate:

    a) the amount of money received by the grocer after selling ALL the boxes of cereal   **[1]**

    b) the total profit made   **[1]**

c) the profit made as a percentage of the cost price [2]

3 A worker at a popular fast food restaurant is paid a basic wage of $600 for a 40-hour week. The overtime paid to a worker is one and a half times the basic wage.

a) Calculate the hourly rate. [1]

b) Calculate the overtime wage if a worker does 8 hours of overtime. [2]

c) Calculate the total wage earned by a worker who works 50 hours in a week. [3]

4 Yashoda lives in Trinidad and is planning a trip to the United States. The current conversion rate is:

US $1.00 = TT $6.30

The limit on her credit card is TT $5000.

a) What is her credit card limit in US dollars? [2]

b) By the last day of her trip she had spent a total of US $650. How many
TT dollars does she have left on her credit card? **[3]**

5 **a)** Anya bought a plot of land for $350 000. After two years she sold the land
for $420 000. Calculate the percentage profit on the sale of the land. **[2]**

b) Daniel purchased a piece of equipment for his workshop for $75 000.
He eventually sold the equipment for $40 000. Calculate the percentage
loss as a result of the sale. **[2]**

c) A new car depreciates in value by 10% per year. Hamlet purchases a new car for $180 000.

i) Calculate the value of the car after one year. [2]

ii) Calculate the value of the car after two years. [2]

6 **a)** Kerry's bank agrees to a loan of $120 000. The loan is to be repaid in five years at a simple interest rate of 10%. Calculate:

i) the total interest to be repaid [2]

ii) the total amount of money to be repaid [2]

iii) his monthly instalment [2]

b) Calculate the compound interest on an investment of $10 000 if it is invested for 5 years at a rate of 2.5% per annum. [2]

7 **a)** Fidellis wants to buy an engagement ring for his girlfriend. The ring he likes is marked at $6500. The store owner gives him a 10% discount. Calculate the amount Fidellis pays for the ring. [2]

b) In Trinidad and Tobago there is a value added tax (VAT) of 15% on items purchased by consumers. The pre-tax price of a camera at the mall is $3000. Calculate the amount of money that a customer would have to pay. [2]

Total Marks _____ / 38

4 Sets

1 The universal set, U, is defined as the set of integers between 4 and 22 inclusive.

A and B are subsets of U such that:

A = {multiples of 3}

B = {odd numbers}

a) State the number of elements in the universal set U.   [1]

b) List the members of the subset A.   [2]

c) List the members of the subset B.   [2]

d) Draw a Venn diagram to represent the relationships among A, B and U.   [3]

2 Sets A and B are defined as follows:

A = {3, 5, 7, 9, 11, 13, 15, 17}

B = {2, 4, 9, 11, 15}

Determine:

a) $n(A)$ [1]

b) $n(B)$ [1]

c) $A \cap B$ [1]

d) $A \cup B$ [1]

3 Use the Venn diagrams to shade the appropriate region.

a) $A \cap B$ [1]

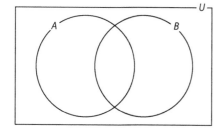

b) $A \cup B$ [1]

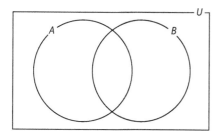

c) A' [1]

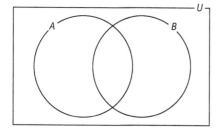

d) $(A \cap B)'$ [1]

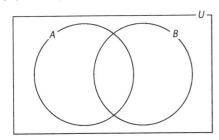

4 If set A = {2, 4, 6}:

   a) state the number of subsets of A [1]

   b) write the subsets of A [2]

5 Classify each of the following as either finite or infinite sets.

   a) even numbers greater than 24 [1]

   b) odd numbers between 25 and 35 [1]

   c) real numbers greater than 4 but less than 18 [1]

   d) multiples of 4 [1]

6 Consider the following sets.

A = ∅

B = {2, 4, 6, 8}

C = {1, 3, 5, 7, …}

D = {4, 6}

E = {1, 4, 7, 9}

a) State which sets are equivalent. [1]

b) State which set is a subset of another. [1]

c) State which two sets are disjoint. [1]

d) State which set is the null set. [1]

e) Give an example of a finite set. [1]

f) Give an example of an infinite set. [1]

g) State $n(B)$. [1]

h) State the number of subsets of B. [1]

7 A class has 36 students. 25 students study Biology (B) and 18 students study Chemistry (C). 2 students study neither Biology nor Chemistry. x students study both Biology and Chemistry.

a) Illustrate this information on a Venn diagram. **[3]**

b) Calculate the value of x. **[2]**

8 There are 40 students in a Form 5 classroom.

23 students study Biology.

15 students study Chemistry.

18 students study Physics.

6 students study Chemistry and Biology.

9 students study Physics and Biology.

5 students study Physics and Chemistry.

x students study Biology, Chemistry and Physics.

2 students study none of the subjects.

a) Illustrate this information on a Venn diagram. **[6]**

b) Write an expression in terms of x for the total number of students in the classroom. **[2]**

c) Determine the number of students who study Biology, Chemistry and Physics. **[2]**

d) Determine the number of students who study Biology only. **[1]**

e) Determine the number of students who study Chemistry and Biology only. **[1]**

Total Marks _____ / 48

1 The diagram below shows a circle of radius 6 cm. AOB = 120°. [Use π = 3.14]

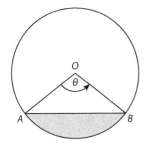

Calculate:

a) the circumference of the circle [2]

b) the area of the circle [2]

c) the area of the minor sector OAB [2]

d) the area of the triangle AOB [2]

e) the area of the shaded region [2]

f) the length of the minor arc AB [2]

g) the length of the major arc AB [2]

h) the perimeter of the shaded region [2]

2 The diagram below, not drawn to scale, shows a glass prism of length 12 cm.

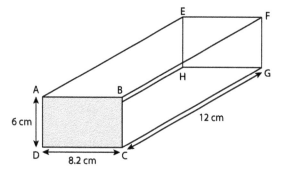

Calculate:

a) the area of the cross-section ABCD [2]

b) the volume of the prism [2]

c) the total surface area, in cm², of the prism   **[4]**

3 Calculate the total area of each of the following shapes. [π = 3.14]

a)   **[3]**

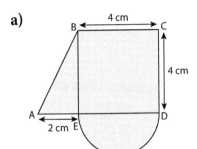

b)   **[4]**

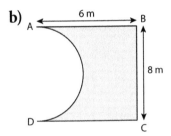

4 The diagram below shows a cylinder. [π = 3.14]

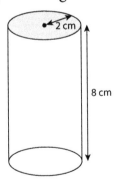

Calculate:

a) the area of the shaded cross-section   **[2]**

b) the volume of the cylinder   **[2]**

c) the area of the curved part of the cylinder   **[2]**

5 A piece of wire is bent to form a square of area 196 cm².

    a) Calculate:

        i)   the length of one side of the square   **[2]**

ii) the perimeter of the square [1]

b) The same wire is now bent to form a circle. $[\pi = \frac{22}{7}]$

   i) State the circumference of the circle. [1]

   ii) Calculate the radius of the circle. [2]

   iii) Calculate the area of the circle. [2]

6 The diagram below, not drawn to scale, shows a new design for a perfume bottle. It consists of a cylinder with a hemisphere on each end.

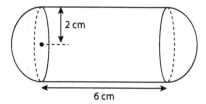

- The curved surface area of a cylinder = $2\pi rh$, where r is the radius of the circular cross-section and h is the height of the cylinder.

- The total surface area of a sphere of radius $r = 4\pi r^2$

- The volume of a sphere of radius $r = \frac{4}{3}\pi r^3$

Assuming $\pi = 3.14$, calculate:

a) the curved surface area of the cylinder [2]

b) the TOTAL surface area of the two hemispheres [2]

c) the TOTAL surface area of the perfume bottle [1]

d) the volume of the cylinder [2]

e) the TOTAL volume of the two hemispheres   **[2]**

f) the TOTAL volume of the perfume bottle   **[1]**

7 The diagram below shows a quadrilateral, not drawn to scale, where AB = BD = BC, angle ABC = 150° and angle BCD = 30°.

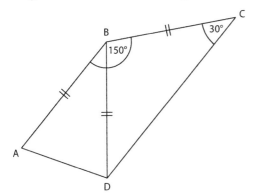

Calculate the size of each of the following angles.

a) angle BDC   **[2]**

b) angle ABD   **[2]**

c) angle ADB [2]

8
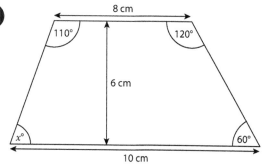

Calculate:

a) the area of the trapezium [2]

b) the value of the angle x [1]

Total Marks _____ / 64

1 A short mathematics quiz was used to test the mental abilities of 30 students in a Form 4 classroom. The results of the test are shown below.

$$
\begin{array}{cccccccccc}
2 & 7 & 2 & 1 & 3 & 6 & 8 & 2 & 3 & 8 \\
4 & 9 & 8 & 5 & 6 & 5 & 4 & 3 & 6 & 6 \\
1 & 6 & 4 & 6 & 1 & 3 & 6 & 6 & 2 & 4
\end{array}
$$

a) Complete the following table.   [5]

Score (x)	Tally	Frequency (f)	$x \times f$
1	III	3	3
2			
3			
4			
5			
6			
7			
8			
9			

b) State the modal score.   [1]

c) State the median score.   [1]

d) Using the table, calculate the mean score.   [3]

e) A student is chosen at random from the Form 4 class. Determine the
   probability that the student scores greater than 5 in the quiz.   [2]

2 A Biology class conducts an experiment to measure the height of a batch of seedlings that have been growing for several days. The data obtained from the experiment are as follows:

Height (cm)	Number of seedlings	Cumulative frequency
1–10	10	
11–20	12	
21–30	22	
31–40	38	
41–50	15	
51–60	3	

a) Complete the table to include the cumulative frequencies. [3]

b) Using an appropriate scale, draw the cumulative frequency graph for the data. Use the graph paper provided on page 38. [5]

c) Use your graph to determine the following:

i) the lower quartile [1]

ii) the median [1]

iii) the upper quartile [1]

iv) the interquartile range [1]

v) the semi-interquartile range [1]

d) Calculate the probability that a seedling chosen at random has a height:

i) less than 35 cm [1]

ii) greater than 42 cm [1]

3 Anya's salary is $6000 per month. She budgets $1000 on rent, $900 on food, $300 on fuel for her car and $800 on utilities. The remainder is used for savings. You are required to illustrate this information in a pie chart.

a) Complete the following table. [4]

Item	Budgeted amount	Angle of sector in pie chart
rent	$1000	
food	$900	
fuel for her car	$300	
utilities	$800	
savings		
Total	$6000	

b) Using a circle of radius 4 cm, construct a pie chart to illustrate the information. [3]

4 In a reggae competition, 10 judges scored a contestant as follows:

$$6 \quad 5 \quad 7 \quad 7 \quad 6 \quad 5 \quad 6 \quad 6 \quad 6 \quad 7$$

Determine:

a) the mean score [2]

b) the median score [2]

c) the modal score [1]

5 The ages of a random sample of 40 students in a lunch room of an international school were recorded. The data collected is shown below.

Age	Frequency
5–7	4
8–10	7
11–13	8
14–16	13
17–19	8

a) In which class interval does the median age lie? [1]

b) In which class interval does the mode lie? [1]

c) State the upper class boundary of the class interval 8–10. [1]

d) State the lower class boundary of the class interval 17–19. [1]

e) State the class width of the class interval 5–7. [1]

f) Draw a histogram to illustrate the distribution of the ages. Use the graph paper provided on page 42. [4]

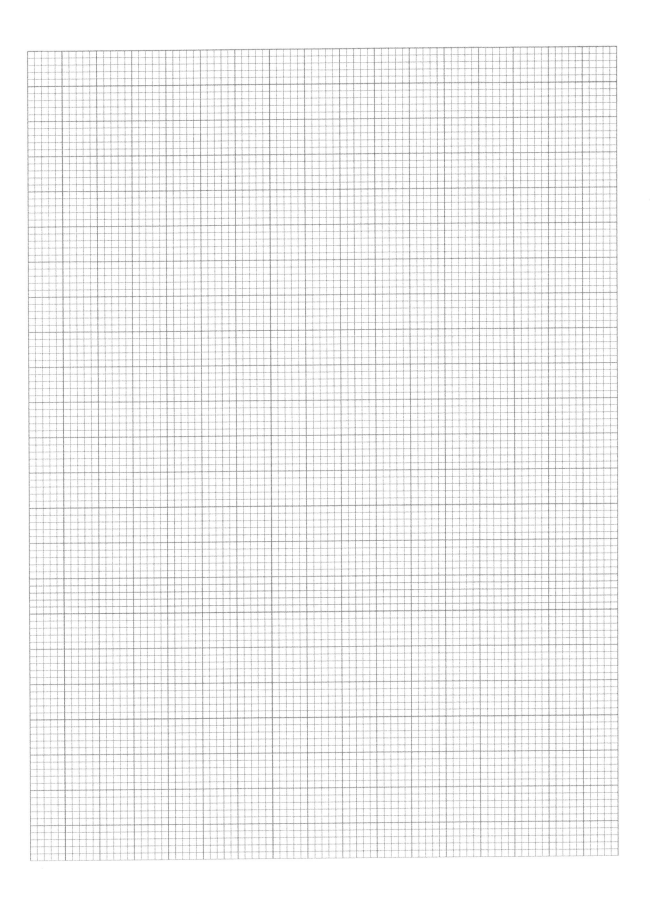

6 **a)** The letters of the word 'MATHEMATICS' are all written on identical pieces of paper and placed in a bowl. The bowl is shaken and a piece of paper is taken from it. What is the probability of drawing the letter M? **[2]**

b) A bag contains 40 coloured balls. The probability of randomly selecting a yellow ball from the bag is $\frac{5}{8}$. How many yellow balls does the bag contain? **[2]**

c) A short quiz is taken in a classroom of 30 students. The scores of the students are as follows:

Score	1	2	3	4	5	6	7	8	9	10
Number of students	2	3	4	6	7	3	1	2	1	1

What is the probability that a student chosen at random scores more than 6? **[2]**

Total Marks _____ / 54

1 **a)** Write as a single fraction in its LOWEST terms.

   i) $\dfrac{x-3}{4} + \dfrac{2x+1}{3}$ **[3]**

   ii) $\dfrac{2x+1}{2} - \dfrac{x-1}{3}$ **[3]**

b) Solve the equation $\dfrac{7x+1}{5} + \dfrac{2x-1}{3} = 4$.

[3]

2 Factorise the following:

a) $4x^2y - 12x^3y^3$

[2]

b) $1 - 9x^2$

[2]

c) $3x^2 + 5x - 2$

[2]

d) $6p^2 + 3pq - 2p - q$

[2]

3 Make x the subject of the formula in EACH of the following:

a) $y = mx + c$ [2]

b) $pq = \dfrac{a}{x}$ [2]

c) $T = k\sqrt{\dfrac{x}{a}}$ [2]

4 Remove the brackets from the following:

a) $3xy(x - 2xy + y)$ [2]

b) $2(x - y) - 3x(2x - 1)$ [2]

c) $\dfrac{x}{3}\left(12x - \dfrac{15}{xy}\right)$ [2]

5 **a)** The binary operation $p * q$ is defined by:

$p * q = (p + q)^2 - 3pq$

Calculate the value of $2 * 3$. [2]

b) If $a = 1$, $b = -2$ and $c = 3$, calculate $\dfrac{ab - c}{b^2 + a}$. [2]

6 **a)** The table below shows the corresponding values of the variables x and y, where y varies directly as x.

x	4	a	7
y	12	9	b

Calculate the values of a and b. [3]

b) The table below shows the corresponding values of the variables x and y, where y varies directly as the square of x.

x	2	a	5
y	16	64	b

Calculate the values of a and b. [3]

c) The table below shows the corresponding values of the variables x and y, where y is inversely proportional to x.

x	2	a	4
y	6	3	b

Calculate the values of a and b. [3]

7 a) Solve this pair of simultaneous equations. [3]

$2x + 3y = 8$

$3x - y = 1$

b) Solve this pair of simultaneous equations.

$y + 2x = 7$

$x^2 - xy = 6$

[6]

8 Solve, for x, the following equations.

a) $2x - 8 = 12$

[2]

b) $\dfrac{2x}{3} + \dfrac{x}{2} = 6$ [3]

c) $8 - 2x < 2$ [2]

9 50 fruit cakes were sold at a cake sale. Of these, x were sold for $50 and the remainder were sold at $35 each.

a) Write an expression, in terms of x, for:

i) the total number of cakes sold at $35 each [1]

ii) the total amount of money collected for the sale of 50 cakes [1]

b) $2200 was collected from the sale of the 50 cakes. Calculate:

i) the number of cakes sold for $50 [3]

ii) the number of cakes sold for $35 [1]

10 a) Daniel works for a fixed salary. Every month he saves $$x$. In the month of December he was given additional duties and was paid extra. He managed to save $400 more than half of his usual amount. How much did Daniel save in December? [2]

b) A piece of rubber tubing of length 42 cm is cut into three pieces. The length of the first piece is x cm. The second piece is 2 cm longer than the first piece. The third piece is three times as long as the first piece.

i) State, in terms of x, the length of EACH of the pieces. [2]

First piece = _____

Second piece = _____

Third piece = _____

ii) Write an expression, in terms of x, to represent the sum of the lengths of the three pieces of rubber tubing. [1]

Sum of the lengths of the pieces = _____

iii) Calculate the value of x. [3]

11 Simplify the following:

a) $x^2 y \times x^5 y^2 \times xy$ [1]

b) $x^5 y^3 \div x^2 y^2$ [1]

c) $\dfrac{ab \times a^2 b^3}{ab}$ [2]

d) $x^0 y^2$ [1]

12 Solve the following quadratic equations by factorising.

a) $x^2 + 7x + 10 = 0$ [2]

b) $3x^2 + 10x + 8 = 0$ [2]

c) $6x^2 - 13x + 5 = 0$ [2]

13 Solve the following quadratic equations using the formula $x = \dfrac{-b \pm \sqrt{b^2 - 4ac}}{2a}$, giving your answers to 2 decimal places.

a) $2x^2 + 5x + 1 = 0$ [4]

b) $x^2 + 7x - 2 = 0$ [4]

14 Factorise the following:

   a) $2 - 18x^2$ [2]

   b) $6x^2 + x - 12$ [2]

   c) $8pr + 12qr - 2ps - 3qs$ [2]

Total Marks _____ / 97

8   Relations, functions and graphs

1 The equation of a line P is given by $y = 2x - 3$.

    a) State the gradient of any line that is parallel to P.   **[1]**

    b) State the gradient of any line that is perpendicular to P.   **[1]**

    c) Determine the equation of a line PARALLEL to P that passes through the point $(2, -3)$.   **[3]**

2 The coordinates of A and B are $(-2, 1)$ and $(3, 11)$. Determine:

    a) the gradient of the line AB   **[1]**

b) the midpoint M of the line AB [2]

c) the length of the line AB [2]

d) the equation of the perpendicular bisector of AB [3]

3 The equation of a straight line is $y = 3x - 6$. The line meets the x-axis at A and the y-axis B. Determine:

a) the coordinates of A [1]

b) the coordinates of B   [1]

c) the area of the triangle OAB, where O is the origin   [1]

4 **a)** Classify EACH of the following relations using one of the following terms.

- one-to-one

- many-to-one

- one-to-many

- many-to-many

i)   [1]

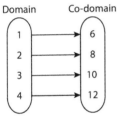

ii)   [1]

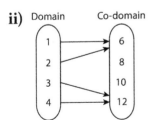

iii)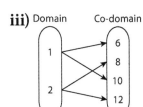

[1]

iv)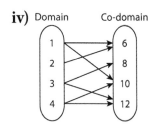

[1]

b) State which relations are considered functions.

[2]

5 The functions f and g are defined as follows:

$f : x \rightarrow 5x - 2$

$g : x \rightarrow \dfrac{1}{3x}$

Calculate:

a) $f(2)$

[1]

b) $f(-1)$

[2]

c) $g(4)$ [1]

d) $fg(x)$ [2]

e) $gf(x)$ [2]

f) $f^{-1}(x)$ [2]

6 Two functions are defined as follows:

$$f(x) = \frac{2x + 1}{x - 3}$$

$$g(x) = 2x + 3$$

a) State the value of x for which $f(x)$ is undefined. [1]

b) Calculate the value of $fg(2)$. [3]

c) Determine $f^{-1}(x)$. [3]

7 Given that $f(x) = 2x^2 + 5x - 3$:

a) write $f(x)$ in the form $f(x) = a(x + b)^2 + c$ where $a, b, c \in \mathbf{R}$ [3]

b) state the equation of the axis of symmetry [1]

c) state the coordinates of the minimum point [1]

d) determine the values of x for which $f(x) = 0$ [2]

e) sketch the graph of $f(x)$ [2]

f) On the graph of $f(x)$ label the following:

  i) the minimum point [1]

  ii) the axis of symmetry [1]

8 The graph below shows the graph $y = f(x)$ where $f(x) = ax^2 + bx + c$.

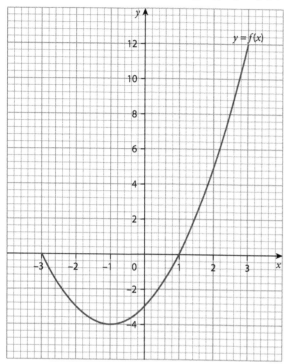

a) State the values of x for which $f(x) = 0$. [2]

b) Determine the equation of the curve. [2]

c) State the value of $f(x)$ when $x = 2$. [2]

d) State the coordinates of the minimum point. [2]

e) State the equation of the axis of symmetry. [2]

f) State the values of x for which $f(x) = -3$. [2]

g) Determine the interval within which x lies when $f(x) < -3$, giving your answer in the form $p < x < q$. [2]

9 **a)** Given $f(x) = x^2 - 2x - 3$, complete the following table and draw the graph. Use the graph paper provided on page 65. [10]

x	-2	-1	0	1	2	3	4	5
$f(x)$								

b) The domain for $f(x)$ is $a \leq x \leq b$. State the value of a and of b. [2]

c) Determine the values of x for which $f(x) = 0$. [2]

d) Determine the coordinates of the turning point on the graph. [1]

e) Determine the gradient of $f(x) = x^2 - 2x - 3$ at the point $x = 2$ by drawing a tangent to the curve at that point. [3]

10 **a)** Given

$$f(x) = 12 - x - x^2$$

and $\quad g(x) = 6 - 2x$

complete the following table. **[6]**

x	−5	−4	−3	−2	−1	0	1	2	3	4
f(x)										
g(x)										

b) On one graph, plot $f(x)$ and $g(x)$. Use the graph paper provided on page 67. **[4]**

c) Use your graph to determine the solution to the following simultaneous equations.

$$y = 12 - x - x^2$$

$$y = 6 - 2x \qquad \textbf{[2]}$$

d) State the values of x for which $12 - x - x^2 = 0$. **[2]**

e) State the range of values for which $f(x) > 0$ in the form $a < x < b$. **[2]**

f) State the range of values of x for which $f(x) < 0$ in the form $x > a$ and $x < b$. **[2]**

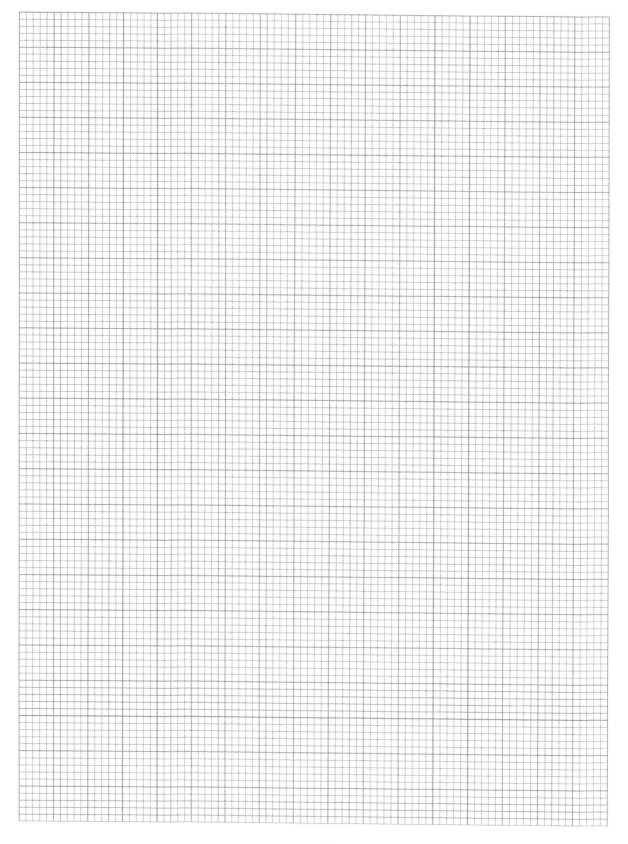

11 The velocity–time graph below shows the motion of an object over a period of 100 seconds.

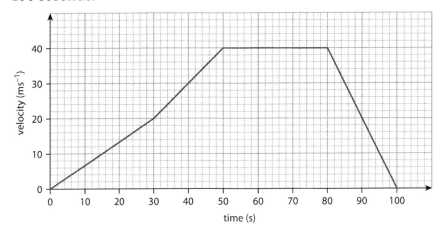

Use the graph to determine the following:

a) the acceleration of the object during the first 30 seconds [2]

b) the acceleration of the object between time $t = 30$ and $t = 50$ seconds [2]

c) the deceleration of the object during the last 20 seconds [2]

d) the acceleration of the object between $t = 50$ and $t = 80$ seconds [1]

e) the distance travelled by the object during the last 50 seconds [2]

12 The diagram below shows the distance–time graph of an old man taking a walk.

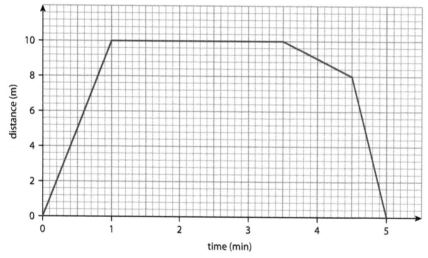

Use the graph to determine the following:

a) the average speed during the first minute of his journey in m/s [2]

b) the length of time for which the old man rested   [1]

c) the average speed during the last 30 seconds   [2]

13 On each of the following graphs, shade the region that satisfies the inequality given.

a) $x \geq 2$ [1]

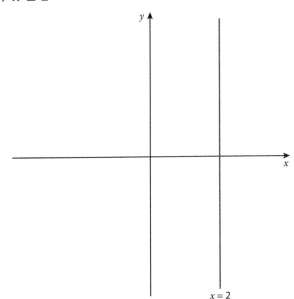

b) $y \geq 4$ [1]

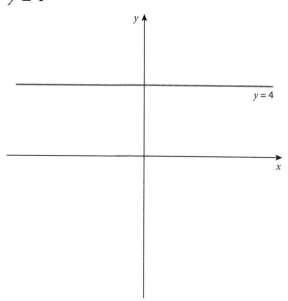

c) $y \leq 2x + 3$ [1]

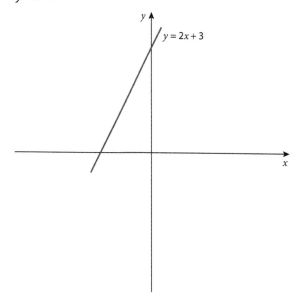

d) $y + 3x \le 7$ [1]

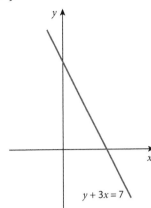

14 A company manufactures two brands of refrigerator, A and B, of similar dimensions. On any given day, the company has to produce at least 4 of refrigerator A and at least 5 of refrigerator B. The workshop is only able to store a total of 15 refrigerators.

a) Write three inequalities that satisfy the conditions identified above.

Inequality 1: _____

Inequality 2: _____

Inequality 3: _____

b) Using the graph paper provided on page 74, draw the lines associated with the inequalities. [4]

c) Shade on your graph the area that satisfies the three inequalities. [2]

d) Refrigerator A is sold for $6000 and refrigerator B is sold for $7000.

  i) Determine the quantities of EACH refrigerator that must be produced to achieve the highest profit.   **[3]**

Quantity of refrigerator A: _____

Quantity of refrigerator B: _____

  ii) Determine the maximum profit that could be achieved.   **[2]**

Maximum profit = _____

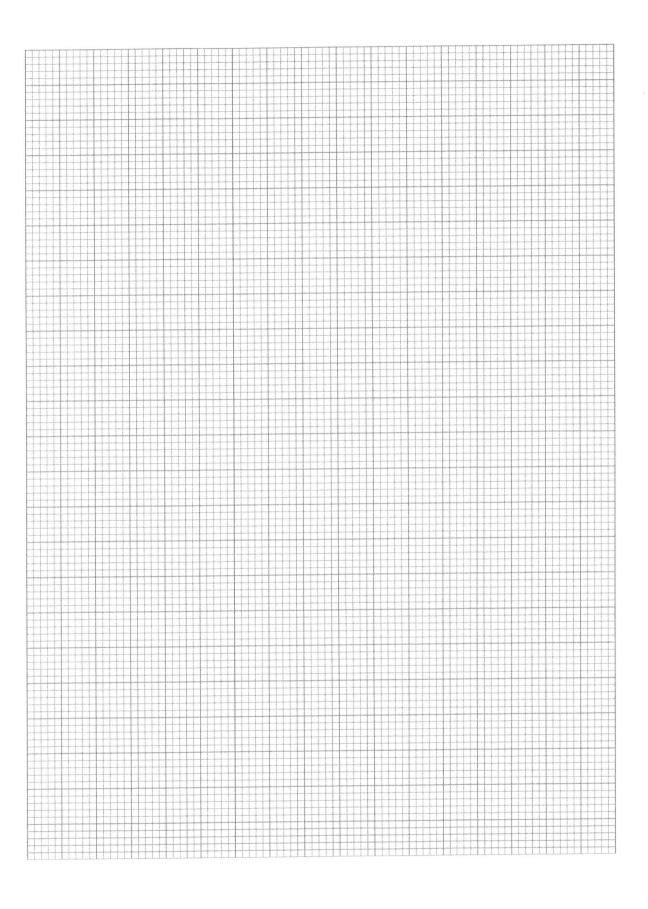

15 The diagram below shows a shaded region bounded by the lines $y = 6 - x$, the x-axis, the y-axis and the line L.

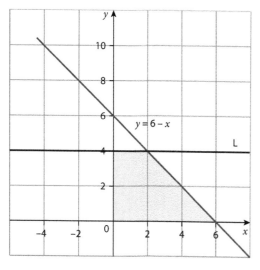

a) Write the equations for:

   i)   the x-axis   [1]

   ii)   the y-axis   [1]

   iii) line L   [1]

b) Write the set of four inequalities that define the shaded region.   [4]

16 The diagram below shows the shaded region bounded by the lines $x + y = 15$, $y = x$, line P and line Q.

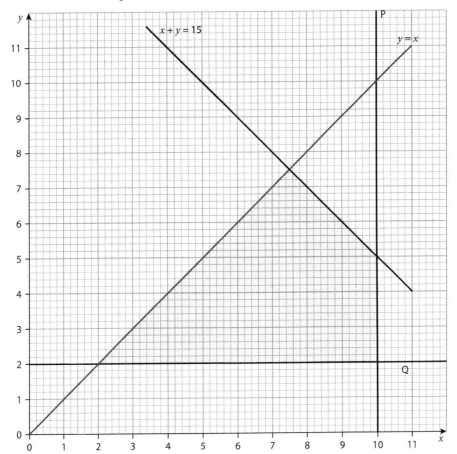

a) Write the equations of lines P and Q. [2]

b) Write the set of four inequalities that define the shaded region. [4]

17 A curve has the following equation: $y = 3 - 5x - 2x^2$.

a) Express the equation in the form $y = a(x + h)^2 + k$. [2]

b) State the equation of the axis of symmetry. [1]

c) State the number of roots for the curve. [1]

d) State the roots of the curve. [2]

e) State whether the curve has a maximum or minimum point. State the coordinates of this point. [2]

18 Determine the inverse of EACH of the following functions.

a) $f(x) = 3x + 2$ [2]

b) $g(x) = 6 - x$ [2]

c) $f(x) = \dfrac{x + 3}{2x - 1}$ [3]

Total Marks _____ / 160

9 Geometry and trigonometry

1 Using a ruler, a pencil and compasses only, construct the following angles.

a) 60° **[3]**

b) 30° **[3]**

c) 45°

[3]

d) 120°

[3]

2 **a)** Using a ruler, a pencil and compasses, construct a triangle ABC with AB = 6 cm, $\widehat{ABC} = 60°$ and $\widehat{CAB} = 90°$. [4]

b) Measure and state the length of AC. [1]

3 Use a ruler, a pencil and compasses only.

a) Construct a triangle ABC in which AB = 7.5 cm, AC = 5.5 cm and angle A = 60°.   [3]

b) Construct the line segment CX which is perpendicular to AB and meets AB at X.   [2]

c) Measure and state the size of the angle BCX.   [1]

4 Use a ruler, a pencil and compasses only.

a) In the space below, construct a triangle ABC in which BC = 8.2 cm, AB = 6.4 cm and AC = 7 cm. **[3]**

b) Construct the lines BD and CD such that ABDC is a parallelogram. **[2]**

5 **a)** Calculate the value of x.

[2]

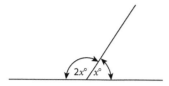

b) Angle ABC is 90°. Calculate the value of x.

[1]

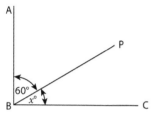

c) Calculate the value of x.

[2]

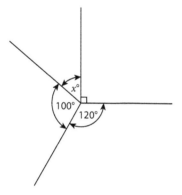

d) Determine the magnitude of the angles a, b, c, d and e, giving reasons for your answers.

[5]

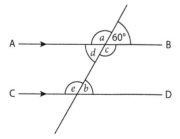

6 Determine θ in each of the following, giving reasons for your answers.

a) [1]

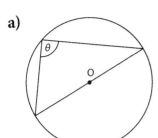

b) [1]

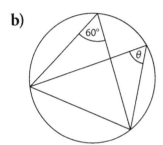

c) [1]

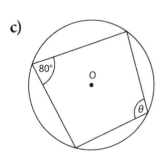

d)

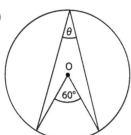

[1]

e)

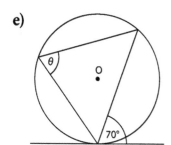

[1]

7 The diagram below, not drawn to scale, shows a circle, centre O. The lines QS and RS are tangents to the circle. The angle QRS = 80°.

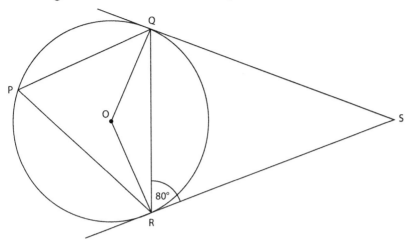

Calculate the following, giving reasons for each step of your answer.

a) angle QPR   [2]

b) angle QOR   [2]

c) angle QSR   [2]

8 The diagram below, not drawn to scale, shows a circle, centre O. PACS is a tangent to the circle. PRQ, ARB and COB are straight lines. Angle RQC = 30°.

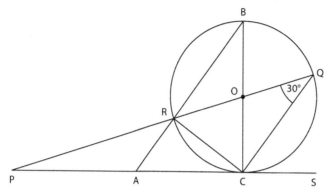

Calculate the following, giving reasons for each step of your answer.

a) angle ROC [2]

b) angle CAB [2]

c) angle QPS [2]

d) angle RCA [2]

9 Calculate the length of AC. **[2]**

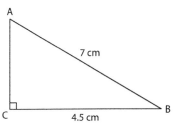

10 Calculate the length of AB. **[2]**

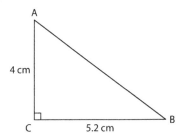

11 Calculate the length RQ.

[2]

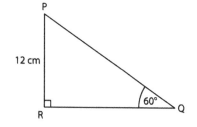

12 Calculate the angle PQR.

[2]

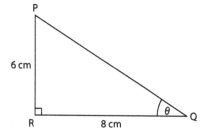

13 Calculate the length RQ. [2]

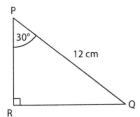

14 Calculate the angle PQR. [2]

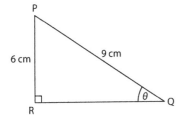

15 Calculate the length RQ.

[2]

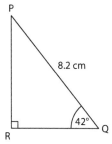

16 Calculate the angle RPQ.

[2]

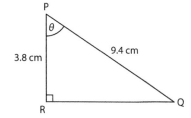

17 Calculate the angle ACB.

[2]

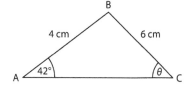

18 Calculate the length BC. [2]

19 Calculate the length BC. [2]

20 Calculate the angle ABC. [2]

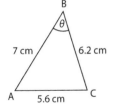

21 The diagram below, not drawn to scale, shows a vertical pole OT standing on a horizontal plane and OPQ are points on the horizontal plane. OT = 11 metres and the angles of elevation of the top of the pole, T, from P and Q are 40° and 30° respectively.

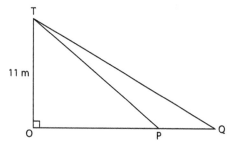

a) Insert the angles of elevation on the diagram. [2]

b) Calculate the length of OP. [2]

c) Calculate the length of PQ. [3]

22 A cargo ship leaves port O to deliver canned fruits to port A. After delivering at port A it heads to port B for another delivery.

The bearing of A from O is 110°.

The bearing of B from A is 030°.

The distance OA is 50 km and the distance OB is 65 km.

a) Sketch a diagram showing the path taken by the cargo ship as it travels from O to A to B.

Label all the bearings and distances given. **[5]**

b) Calculate:

    i) the size of angle OAB **[1]**

    ii) the size of angle OBA **[3]**

iii) the bearing of O from B [1]

23 Triangle A′B′C′ is the image of triangle ABC, under the transformation, T. Describe completely the transformation T. [3]

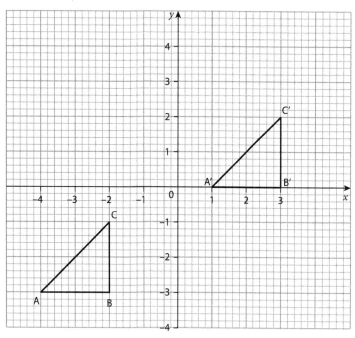

24 Triangle A′B′C′ is the image of triangle ABC, under the transformation, M. Describe completely the transformation M. **[3]**

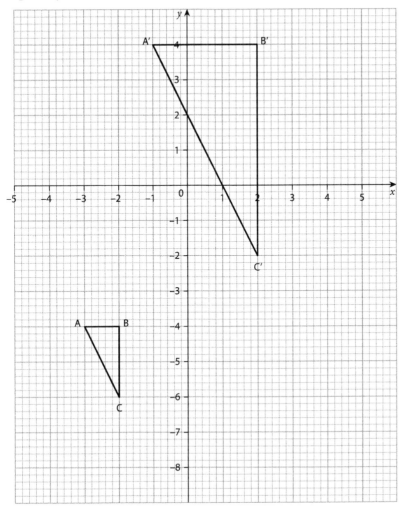

25 Triangle A′B′C′ is the image of triangle ABC, under the transformation, J. Describe completely the transformation J. **[2]**

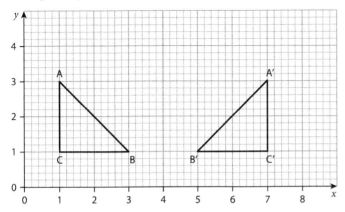

26 The diagram below shows triangle ABC and its image, triangle A′B′C′, after undergoing a rotation.

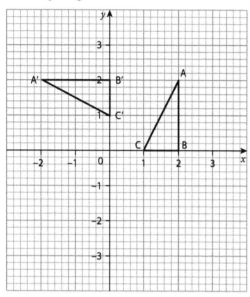

a) Describe the rotation fully by stating:

   i) the centre of rotation [1]

   ii) the angle of rotation [1]

   iii) the direction of rotation [1]

b) State one geometric relationship between triangle ABC and its image. [1]

c) Triangle A′B′C′ is translated by the vector $\begin{pmatrix} 1 \\ -2 \end{pmatrix}$. Determine the coordinates of the image of all the points of the triangle under this transformation. **[3]**

d) Triangle A′B′C′ undergoes a reflection in the line $y = 0$. Draw the image on the graph to show this transformation. **[3]**

27

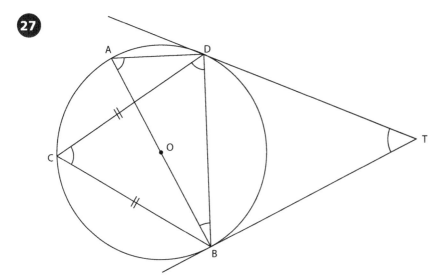

AB is the diameter of the circle BCAD.

BT and DT are tangents to the circle.

CD = CB

Angle BCD = 64°

Calculate, giving reasons for your answers, each of the following:

a) CDB **[2]**

b) BAD [2]

c) ABD [2]

d) DTB [2]

28 A plane travels from point O to point A and then to point B.

OA = 60 km

AB = 80 km

The bearing of A from O is 60°.

The bearing of B from A is 110°.

a) Sketch a diagram showing the journey from O to A to B and label the distances and bearings. **[3]**

b) Calculate the distance OB. **[4]**

c) What is the bearing of O from B? **[4]**

Total Marks _____ / 135

1 Given that A is $\begin{pmatrix} 1 & 3 \\ -2 & 2 \end{pmatrix}$ and B is $\begin{pmatrix} 2 & 1 \\ 1 & -3 \end{pmatrix}$, determine:

a) $A + B$ [2]

b) $A + 2B$ [2]

c) $B - 2A$ [2]

d) AB [2]

e) A^2B [2]

2 Find the determinant of EACH of the following:

a) $\begin{pmatrix} 4 & 3 \\ -2 & 2 \end{pmatrix}$ [1]

Determinant =

b) $\begin{pmatrix} 2 & 3 \\ 5 & 8 \end{pmatrix}$ [1]

Determinant =

c) $\begin{pmatrix} 5 & -3 \\ 1 & 2 \end{pmatrix}$ [1]

Determinant =

d) $\begin{pmatrix} 6 & 3 \\ -1 & -2 \end{pmatrix}$ [1]

Determinant =

3 Find the inverse of EACH of the following:

a) $\begin{pmatrix} 4 & 3 \\ -2 & 2 \end{pmatrix}$ [2]

Inverse =

b) $\begin{pmatrix} 2 & 3 \\ 5 & 8 \end{pmatrix}$ [2]

Inverse =

c) $\begin{pmatrix} 5 & -3 \\ 1 & 2 \end{pmatrix}$ [2]

Inverse =

d) $\begin{pmatrix} 6 & 3 \\ -1 & -2 \end{pmatrix}$ [2]

Inverse =

4 The value of the determinant of $A = \begin{pmatrix} 3 & 6 \\ -2 & x \end{pmatrix}$ is 15.

a) Calculate the value of x. [2]

b) For this value of x, find A^{-1}. **[2]**

c) Show that $A^{-1}A = I$, where I is the identity matrix. **[2]**

5 Given that $B = \begin{pmatrix} 8 & x \\ 4x & 2 \end{pmatrix}$ is a singular matrix. Determine the possible values of x. **[4]**

6 Given the linear equations:

$2x + 3y = 11$

$5x - 2y = -1$

a) write the equations in the form $AX = B$ where A, X and B are matrices **[2]**

$$\begin{pmatrix} & \\ & \end{pmatrix}\begin{pmatrix} \\ \end{pmatrix} = \begin{pmatrix} \\ \end{pmatrix}$$

b) calculate the determinant of the matrix A   [2]

Determinant =

c) determine A^{-1} [2]

$A^{-1} =$

d) Use the matrix A^{-1} to solve for x and y in the simultaneous linear equations.   [5]

7 The points A, B and C have position vectors

$$\overrightarrow{OA} = \begin{pmatrix} 2 \\ 4 \end{pmatrix}$$

$$\overrightarrow{OB} = \begin{pmatrix} 3 \\ 1 \end{pmatrix}$$

$$\overrightarrow{OC} = \begin{pmatrix} 1 \\ -4 \end{pmatrix}$$

relative to an origin O.

Determine:

a) $\overrightarrow{BA}$ [2]

b) $\overrightarrow{BC}$ [2]

c) $\left|\overrightarrow{AB}\right|$ [2]

d) the unit vector in the direction $\overrightarrow{AB}$ [1]

8 Two vectors **a** and **b** are defined as follows:

$\mathbf{a} = 2i + j$

$\mathbf{b} = i - 3j$

Determine:

a) $\mathbf{a} + \mathbf{b}$ [2]

b) $\mathbf{a} - 2\mathbf{b}$ [2]

c) $|\mathbf{a} - \mathbf{b}|$ [3]

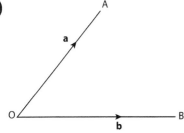

OA and OB are position vectors such that $\overrightarrow{OA}$ = **a** and $\overrightarrow{OB}$ = **b**.

P is the midpoint of OB, and Q is a point on OA such that $\overrightarrow{OQ} = \frac{2}{3}\overrightarrow{OA}$.

a) Show the approximate positions of P and Q on the diagram. [2]

b) Write the following vectors in terms of **a** and **b**.

   i) $\overrightarrow{BA}$ [2]

   ii) $\overrightarrow{QB}$ [2]

   iii) $\overrightarrow{AP}$ [2]

   iv) $\overrightarrow{QP}$ [2]

10 The points A, B and C have position vectors

$$\overrightarrow{OA} = \begin{pmatrix} -2 \\ -4 \end{pmatrix}$$

$$\overrightarrow{OB} = \begin{pmatrix} 1 \\ 5 \end{pmatrix}$$

$$\overrightarrow{OC} = \begin{pmatrix} 3 \\ 11 \end{pmatrix}$$

relative to an origin O.

a) Determine $\overrightarrow{AB}$.   **[2]**

b) Determine $\overrightarrow{BC}$.   **[2]**

c) Use a vector method to show that A, B and C are collinear.   **[3]**

11 The diagram below shows triangle ABC.

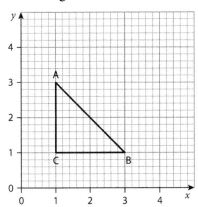

a) State the coordinates of A, B and C. [2]

b) Triangle ABC undergoes two successive transformations, P followed by Q, where
$$P = \begin{pmatrix} 3 & 0 \\ 0 & 3 \end{pmatrix} \text{ and } Q = \begin{pmatrix} -1 & 0 \\ 0 & 1 \end{pmatrix}.$$

    i) State the effect of P on triangle ABC. [2]

ii) Determine the 2×2 matrix that represents the combined transformation of P followed by Q. [3]

iii) Using the matrix in b) ii), determine the coordinates of the image of triangle ABC under this combined transformation. [3]

12 a) Write the 2×2 matrix, A, that represents a reflection in the x-axis. [2]

b) Write the 2×2 matrix, B, that represents a reflection in the y-axis. **[2]**

c) Write the 2×2 matrix, C, that represents a clockwise rotation of $90°$ about the origin. **[2]**

d) Write the 2×1 matrix, D, that represents a translation of -2 units parallel to the x-axis and 4 units parallel to the y-axis. **[1]**

e) The point P (4, 10) undergoes the following combined transformations such that

BD (P) maps P to P$'$

AC (P) maps P to P$''$.

Determine the coordinates of P$'$ and P$''$. **[4]**

13 The diagram below shows triangle OAB. P is the midpoint of AB and Q is the midpoint of OB.

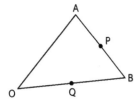

a) Given that $\overrightarrow{OA} = \mathbf{a}$ and $\overrightarrow{OB} = \mathbf{b}$, determine:

   i) $\overrightarrow{AB}$ [2]

   ii) $\overrightarrow{BP}$ [2]

   iii) $\overrightarrow{OP}$ [2]

   iv) $\overrightarrow{AQ}$ [2]

b) The point X lies on OP such that $OX = \frac{3}{5}$ OP. Determine $\overrightarrow{AX}$ in terms of $\mathbf{a}$ and $\mathbf{b}$. [3]

Total Marks _____ / 106